Collins

easy lea

Numbers
bumper book

Ages
3–5

Carol Medcalf

How to use this book

- Find a quiet, comfortable place to work, away from distractions.

- This book has been written in a logical order, so start at the first page and work your way through.

- Try to use the following language as you work through the book together: numbers zero to twenty, next, first, last, before, after, more and less.

- Help with reading the instructions where necessary and ensure that your child understands what to do.

- Children learn and develop at their own rate. If an activity is too difficult for your child, do more of the suggested practical activities (see Parent's tips) and return to the page when you know that they're likely to achieve it.

- Some children find it easier if all the other activities on the page are covered with a blank piece of paper, so only the activity they are working on is visible.

- Always end each activity before your child gets tired so that they will be eager to return next time.

- Help and encourage your child to check their own answers as they complete each activity.

- Let your child return to their favourite pages once they have been completed. Talk about the activities they enjoyed and what they have learned.

Special features of this book:

- **Parent's tip:** situated on every left-hand page, this suggests further activities and encourages discussion about what your child has learned.

- **Progress panel:** situated at the bottom of every right-hand page, the number of stars shows your child how far they have progressed through the book. Once they have completed each double page, ask them to colour in the blank star.

- **Certificate:** the certificate on the last page should be used to reward your child for their effort and achievement. Remember to give your child plenty of praise and encouragement, regardless of how they do.

Published by Collins
An imprint of HarperCollins*Publishers* Ltd
The News Building
1 London Bridge Street
London
SE1 9GF

Browse the complete Collins catalogue at www.collins.co.uk

© HarperCollins*Publishers* Ltd 2018

10 9 8 7 6 5 4 3 2

ISBN 9780008275426

The author asserts the moral right to be identified as the author of this work.

All rights reserved. No part of this publication may be reproduced, stored in a retrieval system, or transmitted, in any form or by any means, electronic, mechanical, photocopying, recording or otherwise, without the prior permission of Collins.

British Library Cataloguing in Publication Data

A Catalogue record for this publication is available from the British Library

All images and illustrations are
© shutterstock.com and
© HarperCollins*Publishers*

Author: Carol Medcalf
Commissioning Editor: Michelle l'Anson
Project Manager: Rebecca Skinner
Cover Design: Sarah Duxbury
Text Design and Layout: Q2A Media
Illustration: Jenny Tulip
Production: Natalia Rebow
Printed in Great Britain by Martins the Printers

Contents

Number search	4	Thirteen, 13	26
Number match	5	Fourteen, 14	27
Number signs	6	Fifteen, 15	28
Zero, 0	7	Sixteen, 16	29
One, 1	8	Seventeen, 17	30
Two, 2	9	Eighteen, 18	31
Three, 3	10	Nineteen, 19	32
Four, 4	11	Twenty, 20	33
Five, 5	12	Which number?	34
Six, 6	13	Odd one out	35
Seven, 7	14	Counting down	36
Eight, 8	15	Missing numbers	37
Nine, 9	16	Colour by numbers	38
Ten, 10	17	Colour by numbers	39
All numbers 0–5	18	Dot-to-dot 1–10	40
All numbers 6–10	19	Dot-to-dot 11–20	41
Odd and even, 0–10	20	Counting in 10s to 100	42
Matching numbers to 10	22	Number hunt	43
Odd one out, 0–10	23	Answers	44
Eleven, 11	24	Certificate	48
Twelve, 12	25		

Number search

Circle each balloon that has a number on it.
Cross (✗) each balloon that has a letter on it.

Colour all the numbers on the phone.

Numbers appear on lots of things around us. Discuss this with your child and start to teach them the difference between numbers and letters so they can differentiate one from the other.

Number match

Look at the numbers.
Draw a line to match each
parcel to the correct door.

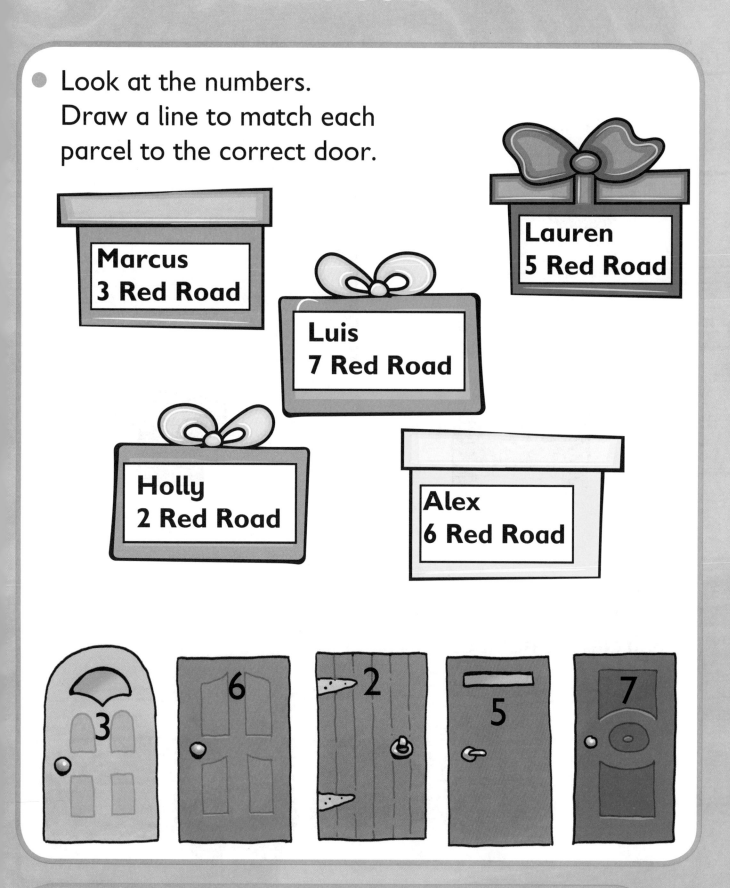

Marcus
3 Red Road

Lauren
5 Red Road

Luis
7 Red Road

Holly
2 Red Road

Alex
6 Red Road

Well done!
Now colour
the star.

Number signs

- Tick (✓) each object that shows a number.

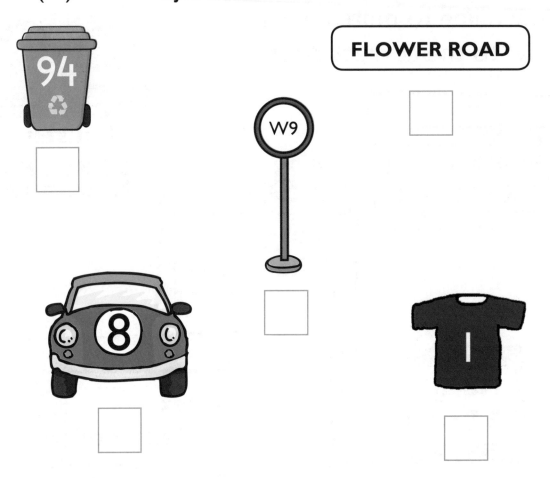

FLOWER ROAD

- Look at the number plates on the cars.
Circle the numbers.

PETE 5

RUTH 22

RAJ 77

Zero, 0

- How many dots on the domino?

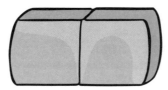

- Start at the red dot and write the number 0.

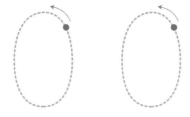

- Circle the squirrels that have no acorns.
  They have 0 acorns.

Well done!
Now colour
the star.

One, 1

- How many dots on the domino?

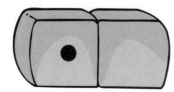

- Start at the red dot and write the number 1.

- Circle each child who has 1 balloon.

Two, 2

- How many dots on the domino?

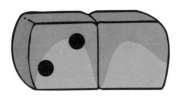

- Start at the red dot and write the number 2.

- Draw wheels on the bikes.
  How many wheels did you need for each bike?

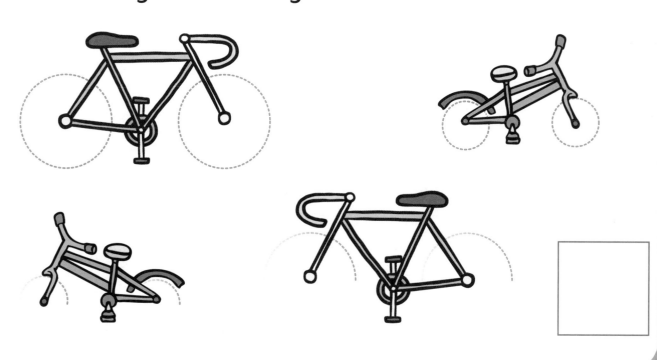

Well done!
Now colour
the star.

Three, 3

How many dots on the domino?

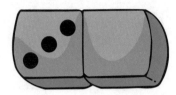

Start at the red dot and write the number 3.

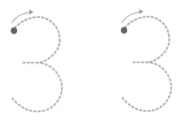

Join the dots to draw triangles.
They all have 3 corners and 3 sides.

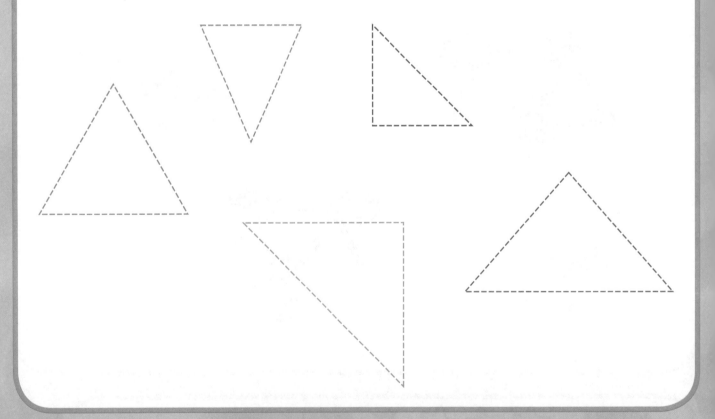

Discuss things that always have the same number, such as 4 legs on a chair, 2 wheels on a bike, 2 eyes, 2 ears, 1 nose and 1 mouth.

10

Four, 4

- How many dots on the domino?

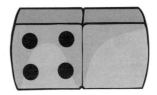

- Start at the red dot and write the number 4.

- Draw 4 legs on each animal.

Well done!
Now colour
the star.

Five, 5

- How many dots on the domino?

- Start at the red dot and write the number 5.

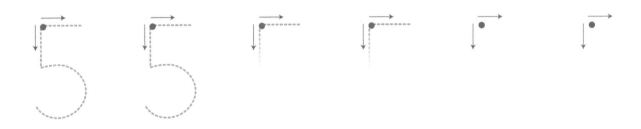

- Colour each star that has 5 points.

Talk about numbers that are important to your child, such as your door number, their age or a sibling's age.
As they get older they may be able to learn your phone number too.

Six, 6

- How many dots on the domino?

- Start at the red dot and write the number 6.

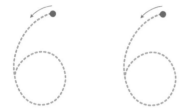

- Draw 6 spots on each butterfly.

Well done!
Now colour
the star.

Seven, 7

- How many dots on the domino?

- Start at the red dot and write the number 7.

- Tick (✓) each picture that shows the number 7.

Look at numbers on everyday objects, such as the ones we have used here. Name the numbers and discuss them with your child.

Eight, 8

- How many dots on the domino?

- Start at the red dot and write the number 8.

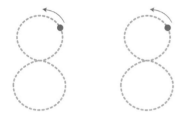

- Draw a line to join each arm or leg of the octopus to a fish.

Well done!
Now colour
the star.

Nine, 9

- How many dots on the domino?

- Start at the red dot and write the number 9.

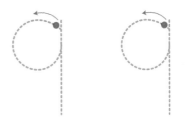

- Draw 9 windows on the house.

It can be fun to cut out a number from card or paper and make a big shadow on the wall, as you would with your fingers. Name all the numbers.

Ten, 10

- How many dots on the domino?

- Start at the red dot and write the number 10.

- This is the first number that you write using two numbers!
Colour the big number 10.

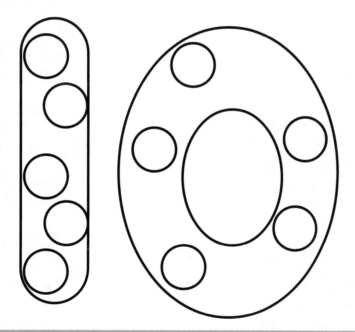

Well done!
Now colour
the star.

All numbers 0–5

● Draw a line from each domino to the dice that shows the same number.

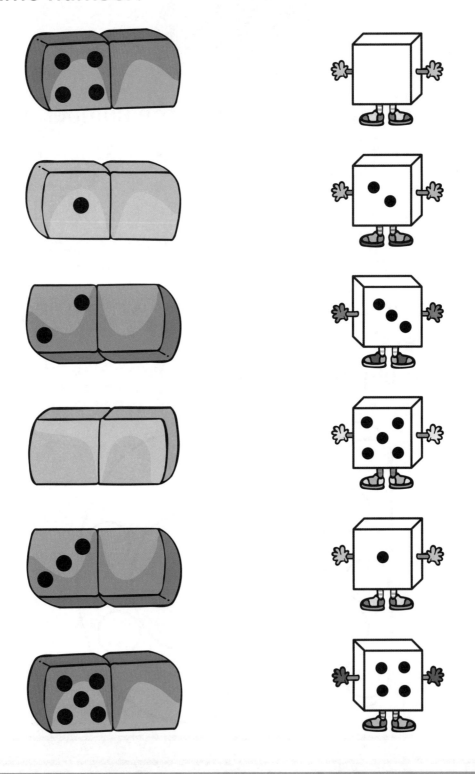

All numbers 6–10

- Draw a line to match each child to the vase that shows the same number.

Well done!
Now colour
the star.

Odd and even, 1–10

Match each set of socks to the correct number.
Colour each set with an odd number of socks red.
Colour each set with an even number of socks blue.

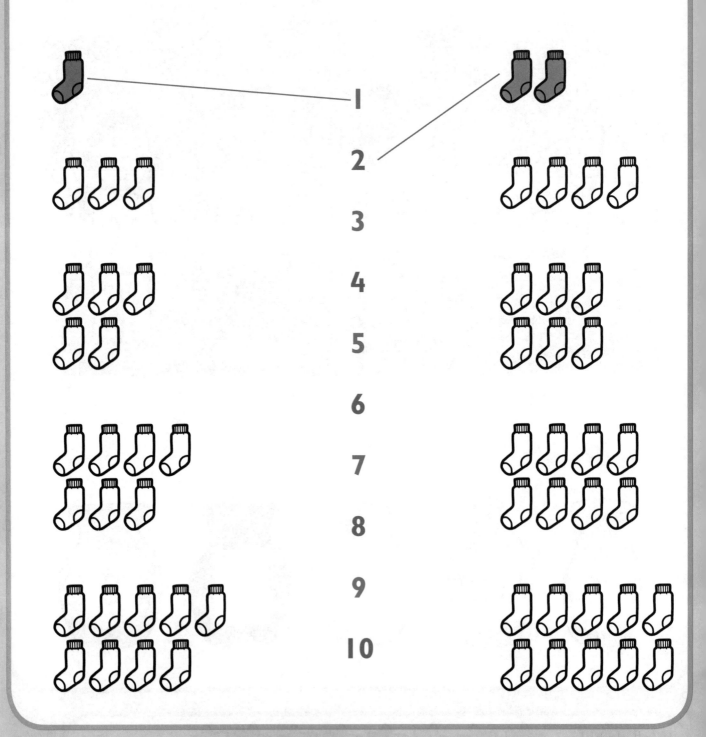

Children are not taught about 'odd' and 'even' numbers until Key Stage 1. However, they will be able to see a pattern and this is the start of learning the concept. Explain that when a sock is alone and not part of a pair, it is 'odd'. This will help with understanding later on.

Odd and even, 1–10

Write the total number of bricks in each picture.
Circle each picture with an odd number of bricks.

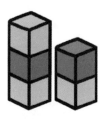 

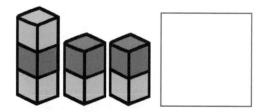

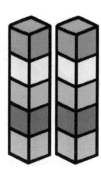 

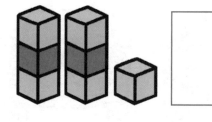

Matching numbers to 10

- Circle the number that is the same as the one at the start of the row.

6	7	9	6	10
8	5	3	2	8
0	1	0	4	9
5	5	7	2	6

- Draw a line to match each car to the correct garage.

Odd one out, 0–10

- Circle the odd one out in each row.

- Colour the cubes with matching numbers the same colour.
Circle the odd one out.

Eleven, 11

● How many dots on the domino?

● Start at the red dot and write the number 11.

● Draw 11 cherries on the cake.

Children love playing with pebbles. Write or paint numbers on pebbles and talk about the numbers. Count out pegs or other small objects to match the number.

Twelve, 12

- How many dots on the domino?

- Start at the red dot and write the number 12.

- It is 12 o'clock.
Draw 2 hands pointing to the number 12 on each clock and watch.

Well done!
Now colour
the star.

Thirteen, 13

- How many dots on the dominoes?

 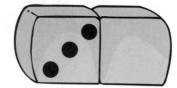

- Start at the red dot and write the number 13.

- 13 bees want to get into the hive.
  Draw a straight line from each bee to the hive.

Fourteen, 14

- How many dots on the dominoes?

- Start at the red dot and write the number 14.

- Draw a ring around each sheep that shows the number 14.
  Copy the number 14 in the box below.

Well done!
Now colour
the star.

Fifteen, 15

- How many dots on the dominoes?

 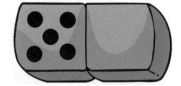

- Start at the red dot and write the number 15.

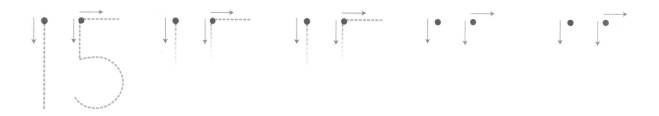

- Draw a line to match each trailer to a tractor with the number 15 on it.

Muffin or fairy cake tins can be used to learn numbers. Write a number at the bottom of each, then count out the correct number of small beads or buttons into each section. As your child grows older, egg boxes and ice cube trays can be used instead.

Sixteen, 16

- How many dots on the dominoes?

- Start at the red dot and write the number 16.

16

- Colour each section with a number 16 to reveal a picture.

Seventeen, 17

- ## How many dots on the dominoes?

- ## Start at the red dot and write the number 17.

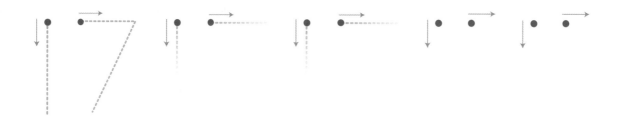

- ## Draw a line from the fishing rod to fish number 17.

Eighteen, 18

- How many dots on the dominoes?

- Start at the red dot and write the number 18.

- Take the mouse through the maze by following the numbers in order.

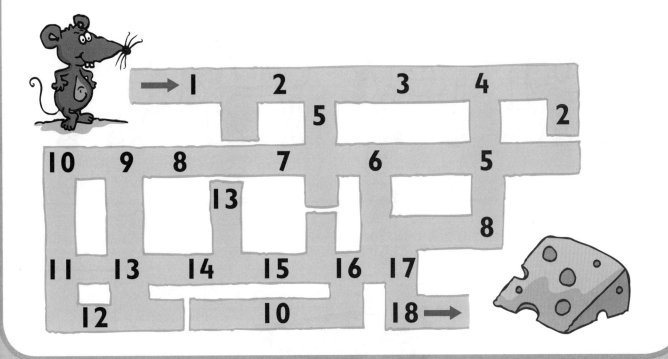

Well done!
Now colour
the star.

Nineteen, 19

- How many dots on the dominoes?

- Start at the red dot and write the number 19.

- Which horse will get the carrot?
  Write the answer in the box.

Use a hole punch to put 20 holes around the edge of a paper plate. Number the holes 1–20. Using a piece of wool or cotton, your child can sew in and out of each hole in order.

Twenty, 20

- How many dots on the dominoes?

- Start at the red dot and write the number 20.

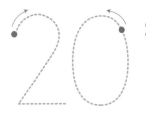

- 20 fingers and toes!
  Colour the nails.

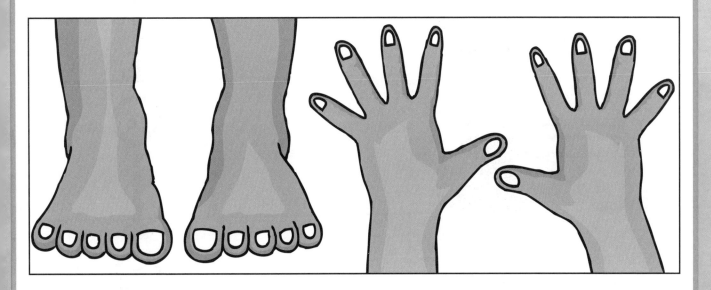

Well done!
Now colour
the star.

Which number?

- Circle the number that is the same as the one at the start of the row.

10	12	10	15	13
11	17	11	16	18
19	11	18	19	14
14	14	20	18	12

- Draw a line to match each bus to the correct bus stop.

Lucky dip: using a small box, some numbers and shredded paper, hide numbers in the paper and then take it in turns to pull one out. Name the number and count out the same number of objects from a supply of bricks or similar.

Odd one out

- Circle the odd one out in each row.

- Colour the cubes that are the same to make them match. Circle the odd one out.

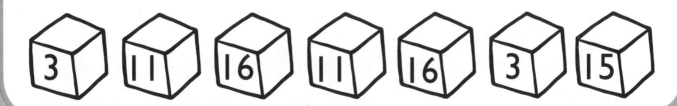

Counting down

Help the rocket to blast to the moon.
Start at number 10 and count backwards to 0.
Colour the circles.

It is great for children to practise counting backwards and the ideal way to start is:
10, 9, 8, 7, 6, 5, 4, 3, 2, 1, blast off!

Missing numbers

Fill in the missing numbers.

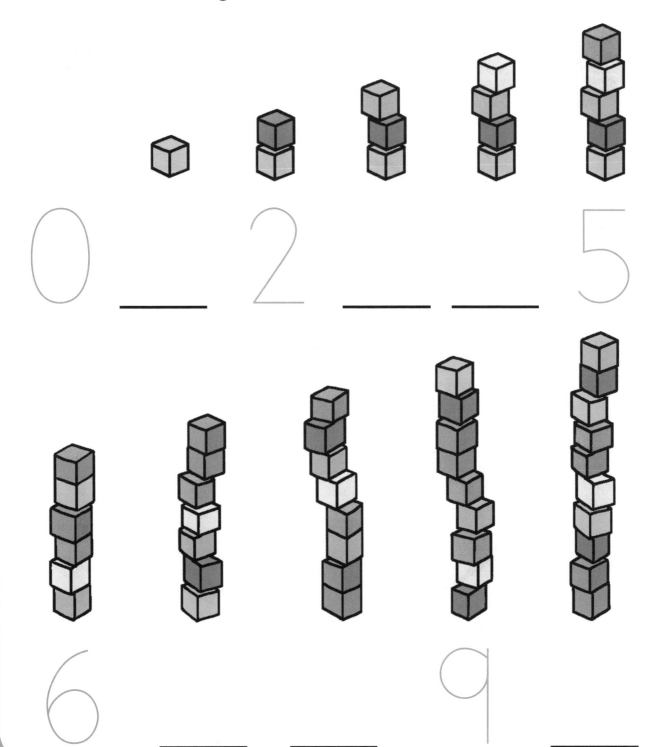

0 __ 2 __ __ 5

6 __ __ __ 9

Well done!
Now colour
the star.

Colour by numbers

Colour the picture using the colours below.

If you have lots of toys cars, label a few by putting numbers on their roofs, one on each car. Using a large sheet of card from an old box, make a car park by drawing 'parking spaces' on the card. Number each space and then ask your child to match each car to the correct space.

Colour by numbers

Colour the picture using the colours below.

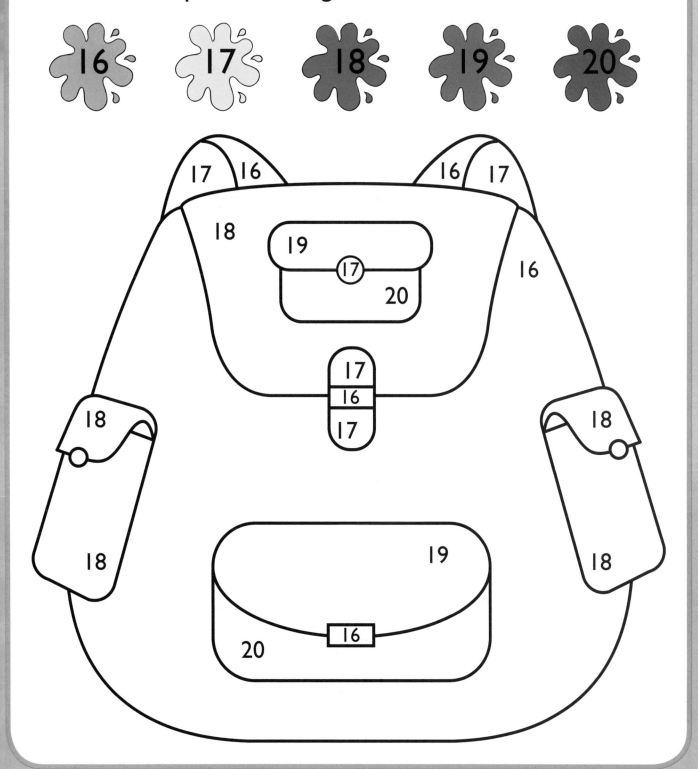

Well done!
Now colour the star.

Dot-to-dot 1–10

- Start at number 1.
  Join the dots in the correct order to make a picture to colour.

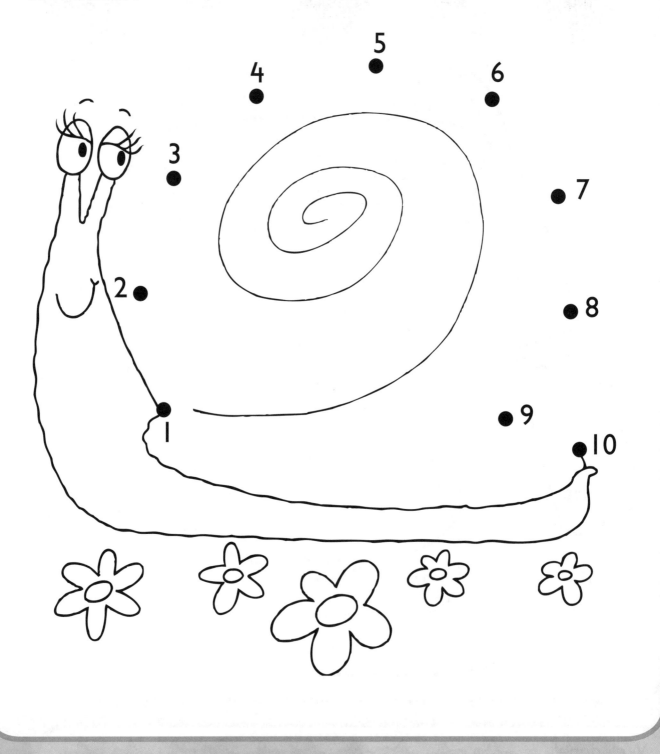

Make cards numbered 0–10 for your child to sequence and play with. When joining the dots in order, having cards with the numbers in the correct order above will help them if needed.

Dot-to-dot 11–20

- Start at number 11.
  Join the dots in the correct order to make a picture.

Well done!
Now colour
the star.

Counting in 10s to 100

- Write the numbers 1 to 10 in order.

10

20

30

40

50

60

70

80

90

100

Children are not expected to count to 100 or count in 10s until Year 1, but some younger children want to learn numbers beyond 20 and enjoy saying 10, 20, 30, etc. Introduce these pages to them and teach them according to their individual ability.

Number hunt

- The grid shows the numbers from 1 to 100.
  Look for numbers around your house, e.g.
  on food packets, birthday cards and remote controls.
  Colour the numbers that you find on the grid.

1	2	3	4	5	6	7	8	9	10
11	12	13	14	15	16	17	18	19	20
21	22	23	24	25	26	27	28	29	30
31	32	33	34	35	36	37	38	39	40
41	42	43	44	45	46	47	48	49	50
51	52	53	54	55	56	57	58	59	60
61	62	63	64	65	66	67	68	69	70
71	72	73	74	75	76	77	78	79	80
81	82	83	84	85	86	87	88	89	90
91	92	93	94	95	96	97	98	99	100

Well done!
Now colour
the star.

Answers

Page 4

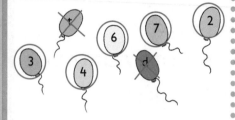

The buttons with numbers on can be any colour.

Page 5

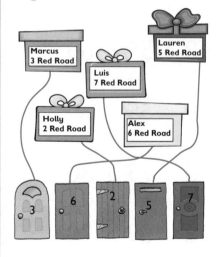

Page 6

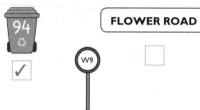

Page 7

Page 8

Page 9

2

Page 10

Page 11

Page 12

The stars with five points can be any colour.

Answers

Page 13

The six spots on each butterfly can be any colour and in any position.

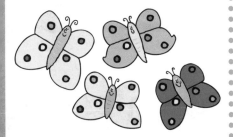

Page 14

Page 15

Each arm or leg of the octopus can be joined to any fish.

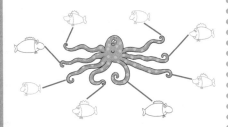

Page 16

Page 17

The number 10 can be any colour.

Page 18

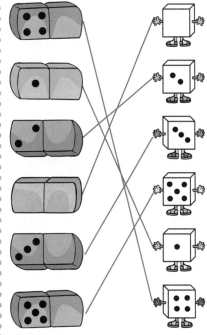

Page 19

Page 20

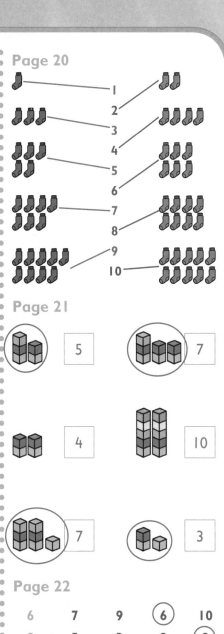

Page 21

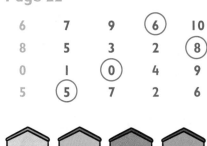

Page 22

6	7	9	⑥	10
8	5	3	2	⑧
0	1	⓪	4	9
5	⑤	7	2	6

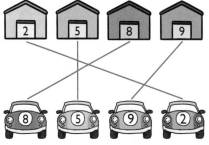

Answers

Page 23

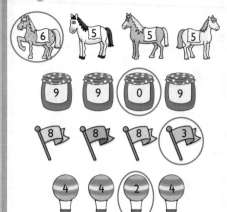

Any colour can be used for each matching pair of cubes.

Page 24

11 cherries drawn, e.g.

Page 25

Page 26

Page 27

14

Page 28

Page 29

The sections containing the number 16 can be any colour.

Page 30

Page 31

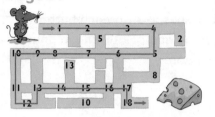

Page 32

19

Page 33

Each nail can be any colour.

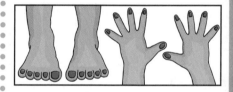

Page 34

10	12	(10)	15	13
11	17	(11)	16	18
19	11	18	(19)	14
14	(14)	20	18	12

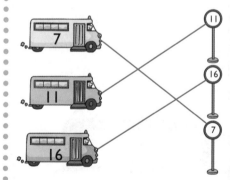

Answers

Page 35

Any colour can be used for each matching pair of cubes.

Page 36

Any colours can be used for the circles.

Page 37

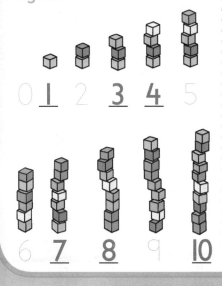

Page 38

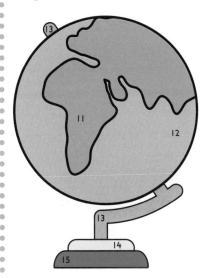

Page 39

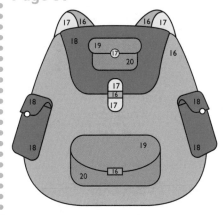

Page 40

Any colours can be used.

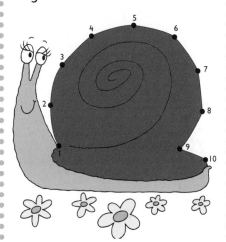

Page 41

Page 42

10 ·········   20 ·········
30 ·········   40 ·········
50 ·········   60 ·········
70 ·········   80 ·········
90 ·········   100 ·········

Page 43

Numbers coloured to show those found around the house.

Collins Easy Learning

Certificate of Achievement

Well Done!

This certificate is awarded to ....

for successfully completing ....

Age ....

Date ....

Signed ....

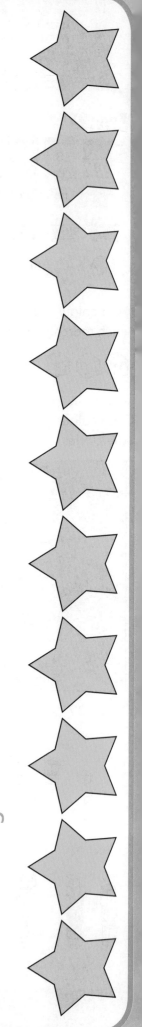